BIZARRE BEAST B

AFRICANIZED HONEYBEE VS. ARMY ANT

Gareth Stevens
PUBLISHING

By Therese Shea

Please visit our website, www.garethstevens.com. For a free color catalog of all our high-quality books, call toll free 1-800-542-2595 or fax 1-877-542-2596.

Cataloging-in-Publication Data

Names: Shea, Therese.
Title: Africanized honeybee vs. army ant / Therese Shea.
Description: New York : Gareth Stevens Publishing, 2019. | Series: Bizarre beast battles | Includes glossary and index.
Identifiers: LCCN ISBN 9781538219232 (pbk.) | ISBN 9781538219256 (library bound) | ISBN 9781538219249 (6 pack)
Subjects: LCSH: Africanized honeybee–Juvenile literature. | Army ants–Juvenile literature.
Classification: LCC QL568.A6 S43 2019 | DDC 595.79'9–dc23

First Edition

Published in 2019 by
Gareth Stevens Publishing
111 East 14th Street, Suite 349
New York, NY 10003

Designer: Katelyn E. Reynolds
Editor: Monika Davies

Photo credits: Cover, p. 1 (Africanized honeybee) aeiddam0853578919/Shutterstock.com; cover, p. 1 (army ant) Dr Morley Read/Shutterstock.com; cover, pp. 1–24 (background texture) Apostrophe/Shutterstock.com; pp. 4–21 (bee and ant icons) Kaimen/Shutterstock.com; p. 5 Felipe Duran/Shutterstock.com; p. 7 Konrad Wothe/naturepl.com/Nature Picture Library/ Getty Images; p. 8 Heidi & Hans-Juergen Koch/Minden Pictures/Getty Images; pp. 9, 17, 19, 21 (army ant) Mark Moffett/ Minden Pictures/Getty Images; pp. 10, 18 Lian van den Heever/Gallo Images/Getty Images; p. 11 Konrad Wothe/Minden Pictures/Getty Images; p. 12 mikeledray/Shutterstock.com; p. 13 Patrick K. Campbell/Shutterstock.com; p. 14 FLAVIO CONCEIÇÃO FOTOS/Moment Open/Getty Images; p. 15 Konrad Wothe/LOOK-foto/Getty Images; p. 16 Rodney Mehring/ Shutterstock.com; p. 21 (Africanized honeybee) SCOTT CAMAZINE/Science Source/Getty Images.

Printed in the United States of America

CPSIA compliance information: Batch #CS18GS: For further information contact Gareth Stevens, New York, New York at 1-800-542-2595.

CONTENTS

Killer Bees 4

Ants on the March 6

Colossal Colonies 8

Small vs. Smaller 10

Adaptations for Attack 12

Vicious Venom 14

Why So Angry? 16

Bug Grub 18

The Winner? 20

Glossary 22

For More Information 23

Index 24

Words in the glossary appear in **bold** type the first time they are used in the text.

KILLER BEES

Africanized honeybees are known by another—scarier—name: killer bees! They're called this because of the **aggressive** way they guard their hive. Their stings have killed both people and animals. In 1956, a scientist from Brazil brought African honeybees to South America. He hoped they'd produce more honey. Some escaped and **mated** with wild European honeybees. The honeybees that resulted were named Africanized honeybees. They're now found throughout the southern United States, and they continue to move north.

5

Though Africanized honeybees are dangerous, there's a greater chance you'll be struck by lightning than killed by them!

ANTS ON THE MARCH

One ant doesn't strike fear in most people, but a whole army of ants might! "Army ants" are ants that swarm, or move in a large group, to look for food instead of building a **permanent** nest.

Perhaps the most famous species, or kind, of army ant is *Eciton burchellii*, or *E. burchellii*. These ants live in Central and South America. When army ants march through the rain forests, they scare creatures from their hiding places. This provides other animals with prey to eat.

7

E. BURCHELLII ARMY ANTS AREN'T USUALLY DANGEROUS. HOWEVER, THEY CAN BE A RISK TO PEOPLE WHO ARE **ALLERGIC** TO ANT STINGS.

COLOSSAL COLONIES

Imagine what would happen if Africanized honeybees and *E. burchellii* army ants battled each other in an **insect** war! Let's see how they match up.

NUMBER OF AFRICANIZED HONEYBEES IN A COLONY:
20,000 TO 90,000

First, they're both known for living in large colonies. Africanized honeybee queens can lay around 1,500 eggs in a day. An *E. burchellii* ant queen lays up to 100,000 eggs in a 3-week period. The army ants outnumber the honeybees. However, this is just one fight in the beast battle!

9

SMALL VS. SMALLER

An army ant colony is larger than an Africanized honeybee colony. But would that matter if the honeybees themselves are larger? Let's compare the size of these insects' bodies.

LENGTH OF AFRICANIZED HONEYBEE:
ABOUT 0.75 INCH (19 mm)

E. burchellii army ants can be different sizes. Some are as tiny as 0.12 inch (3 mm). The smallest ants collect food, while the largest guard the colony. The killer bees are larger. They win this part of the battle, but both bugs are pretty small!

LENGTH OF E. BURCHELLII ARMY ANT: UP TO 0.47 INCH (12 mm)

ADAPTATIONS FOR ATTACK

Both Africanized honeybees and *E. burchellii* army ants have **adaptations** that help them fight their enemies. They would certainly use these in their beast battle with each other!

ADAPTATIONS OF THE AFRICANIZED HONEYBEE:
- WINGS TO FLY THROUGH THE AIR
- STINGER
- VENOM

The honeybees can fly around 15 miles (24 km) per hour. However, the army ants are fast for their size. They travel up to 66 feet (20 m) per hour. Both animals have stingers and venom, but the army ant can rip small creatures apart with their mandibles. Nasty!

ADAPTATIONS OF THE E. BURCHELLII ARMY ANT:
- LONG, POINTED **MANDIBLES** FOR SEIZING AND CUTTING UP PREY
- STINGER
- VENOM

AFRICANIZED HONEYBEE VENOM:
JUST AS POWERFUL AS EUROPEAN HONEYBEE
VENOM, BUT INJECTED IN SMALLER DOSES

VICIOUS VENOM

Both insects **inject** venom into other creatures to guard their colonies or attack prey. They use a stinger on the rear of their body. How deadly is this venom?

One Africanized honeybee's venom isn't very powerful. However, these bees attack in huge swarms. Multiple bee stings can kill a human! The venom of a whole band of army ants can quickly stop and kill a larger creature. This matchup seems like a tie! What do you think?

E. BURCHELLII ARMY ANT VENOM: WORKS QUICKLY TO KILL AND BREAK DOWN A CREATURE'S INSIDES

AFRICANIZED HONEYBEE SWARM:
- OVER 1,000 BEES MAY ATTACK TO GUARD THEIR HIVE
- FOLLOW AN ENEMY AS FAR AS 1/4 MILE (0.4 km)
- REMAIN AGGRESSIVE FOR UP TO 24 HOURS

WHY SO ANGRY?

Both Africanized honeybees and E. burchellii army ants are frightening in great numbers. Unfortunately for their enemies, that's often the way they travel—and attack.

The E. *burchellii* army ants move the location of their nest often—and actually create their nest using their bodies! But the colony needs food. Watch out when they're on the march! Africanized honeybees can hold a **grudge**, but the army ants' path of prey is famous. Each group of creatures is scary for its own reasons.

E. BURCHELLII ARMY ANT SWARM:

- AS MANY AS 500,000 ANTS MAY BE SENT OUT TO FIND FOOD
- CAPTURE ABOUT 30,000 PREY IN ONE DAY
- WORK TOGETHER IN TEAMS TO CARRY PREY BACK TO NEST

BUG GRUB

Each of these alarming insects hungers for certain kinds of food. Could this have an effect on the beast battle?

AFRICANIZED HONEYBEES EAT:
- POLLEN
- NECTAR
- HONEY (PARTIALLY **DIGESTED** NECTAR)

E. burchellii army ants are carnivores, which means they're meat eaters. They'd have no problem attacking and eating a whole swarm of killer bees in their path—it they could catch the flying insects! Could the bees fight off such an attack?

E. BURCHELLII ARMY ANTS EAT:

- INSECTS SUCH AS WASPS, ANTS, COCKROACHES, GRASSHOPPERS, AND BEETLES
- **ARACHNIDS** SUCH AS TARANTULAS AND SCORPIONS

THE WINNER?

Now you know some ways these two bugs can be compared. An *E. burchellii* army ant colony is larger than an Africanized honeybee colony, but the honeybees have the larger body. The honeybees are faster, but ants can chop up prey with their mandibles. Both have venomous stingers. The army ants look for animals to eat, but the honeybees can get very angry.

If these insects came face to face, who'd win? You have the facts. Now decide!

Now that you know more about these two amazing insects, you can imagine who'd win this bizarre beast battle and why!

GLOSSARY

adaptation: a change in a type of animal that makes it better able to live in its surroundings

aggressive: showing a readiness to attack

allergic: to have a sensitivity to usually harmless things in the surroundings, such as dust, pollen, or insect stings

arachnid: one of a large class of small animals that includes spiders, scorpions, ticks, daddy longlegs, and mites

digest: to break down food inside the body so that the body can use it

grudge: a strong feeling of anger toward someone or something that lasts a long time

inject: to use sharp teeth to force venom into an animal's body

insect: a small, often winged, animal with six legs and three main body parts

mandible: a mouthpart of insects or arachnids used to bite or hold food

mate: to come together to make babies

permanent: meant to last a long time

venom: something an animal makes in its body that can harm other animals

FOR MORE INFORMATION

BOOKS

Owings, Lisa. *Killer Bees*. Minneapolis, MN: Bellwether Media, 2013.

Pearson, Scott. *Africanized Honeybees*. North Mankato, MN: Black Rabbit Books, 2017.

Twist, Clint. *The Life Cycle of Army Ants*. Mankato, MN: NewForest Press, 2013.

WEBSITES

Africanized Honey Bees
utahpests.usu.edu/uppdl/files-ou/factsheet/africanized-bees.pdf
Discover more about the Africanized honeybee with this Utah Pests fact sheet.

World's Deadliest: Army Ants Eat Everything
video.nationalgeographic.com/video/worlds-deadliest-ngs/deadliest-army-ants
Watch National Geographic's amazing video about army ants.

INDEX

adaptations 12, 13

aggression 4, 16

beetles 19

Brazil 4

carnivores 19

Central America 6

cockroaches 19

colonies 8, 9, 10, 11, 14, 17, 20

eggs 9

European honeybees 4, 14

grasshoppers 19

hive 4, 16

honey 18

mandibles 13, 20

nectar 18

nest 6, 17

pollen 18

prey 6, 13, 14, 17, 20

queens 9

scorpions 19

South America 4, 6

stinger 12, 13, 14, 20

swarms 6, 15, 16, 17, 19

tarantulas 19

United States 4

venom 12, 13, 14, 15, 20

wasps 19

wings 12